MW00446206

MOLECULES & ELEMENTS
Science for Kids
Children's Chemistry Books Edition

SPEEDY
PUBLISHING

Speedy Publishing LLC
40 E. Main St. #1156
Newark, DE 19711
www.speedypublishing.com

Copyright 2015

All Rights reserved. No part of this book may be reproduced or
used in any way or form or by any means whether electronic or
mechanical, this means that you cannot record or photocopy
any material ideas or tips that are provided in this book

A molecule is formed
when atoms of the
same or different
elements combine.

A molecule is the smallest amount of a chemical substance that can exist.

Molecules are made up of atoms that are held together by chemical bonds.

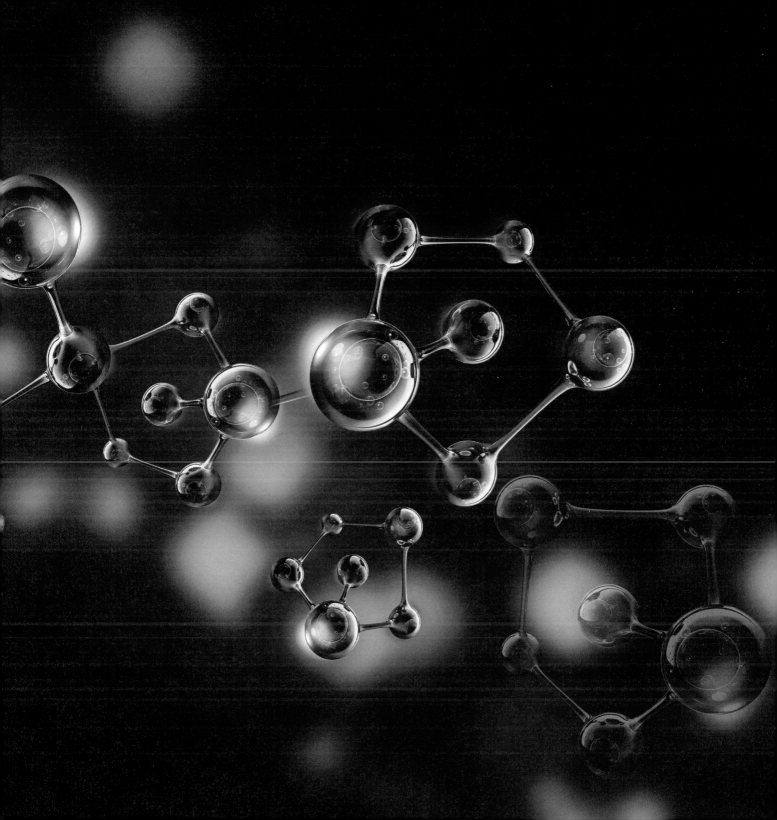

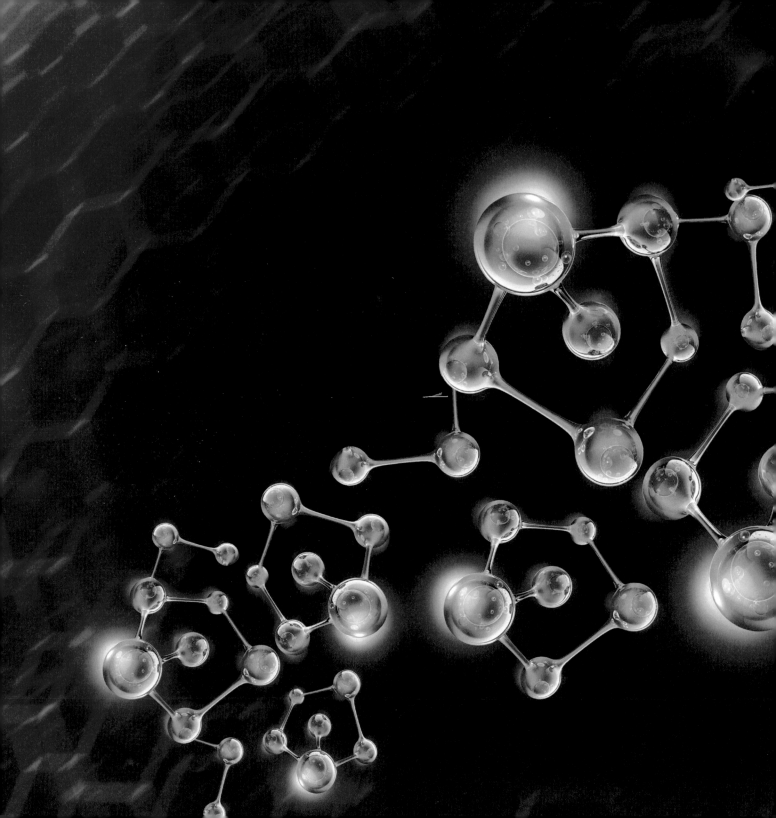

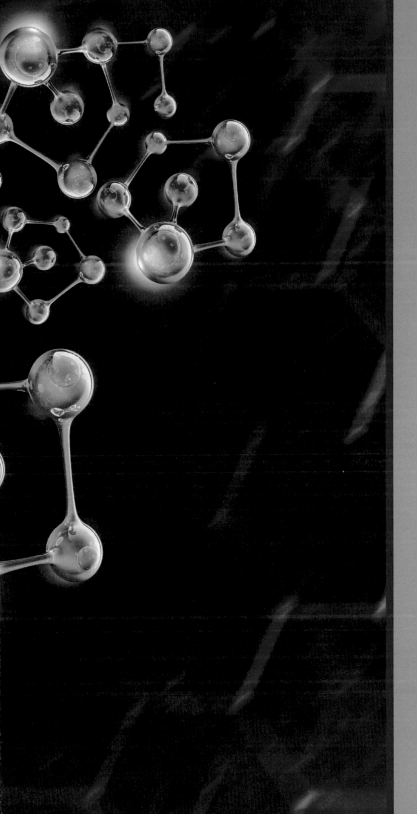

All the stuff
around you
is made up of
molecules.

Molecules can vary greatly in size and complexity.

Molecules can have different shapes. Some are long spirals while others may be pyramid shaped.

A human body is made up of trillions and trillions of different types of molecules.

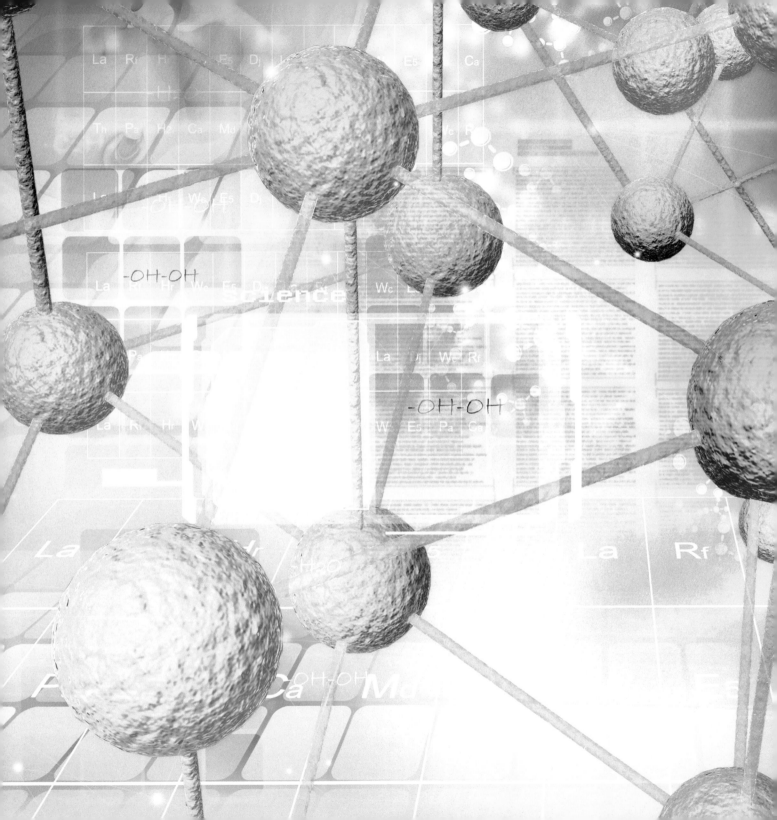

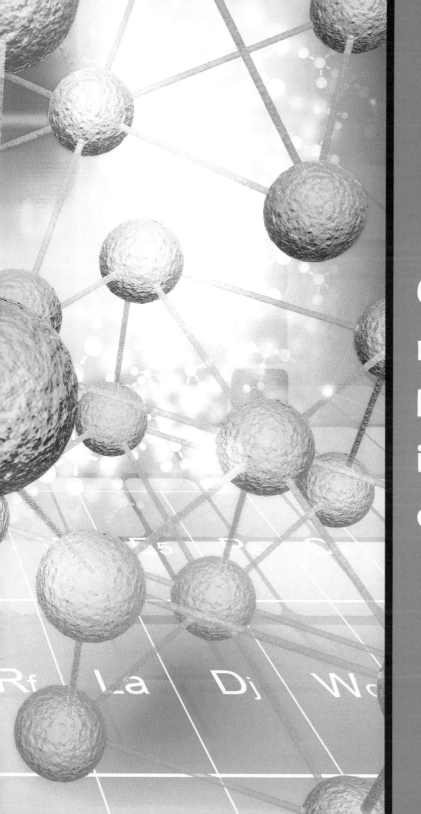

66% of the mass of the human body is made up of oxygen atoms.

Most molecules are far too small to be seen with the naked eye.

The mass of a molecule is called the molecular mass. It is worked out by adding the mass of all the atoms in it.

An element is a pure substance that is made from a single type of atom.

CONH$_2$

CONH

Te

127,60

Sb

121,76

Sn

118,71

84

Po

83

Bi

208,98

82

Pb

207,2

115

Uup

N

38

Cr	Mn	Fe	Co	Ni	Cu
				28	
			58.933	58.693	63.546

42	**43**	**44**	**45**	Palladium	Silver
Mo	Tc	Ru	Rh	**46**	**47**
		101.07(2)	102.91	Pd	Ag
				106.42	107.87

75	Osmium	Iridium	Platinum	Gold
Re	**76**	**77**	**78**	**79**
	Os	Ir	Pt	Au
	190.23(2)	192.22	195.08	196.97

Bohrium	Hassium	Meitnerium	Darmstadtium	Roentgenium	Copern
107	**108**	**109**	**110**	**111**	**112**
Bh	Hs	Mt	Ds	Rg	Cn
[270]	[277.15]	[276.15]	[281.16]	[280.16]	[285.17]

Samarium	Europium	Gadolinium
62	**63**	
Sm		

Elements are the building blocks for all the rest of the matter in the world.

There are currently 118 known elements. Of these, only 94 are thought to naturally exist on Earth.

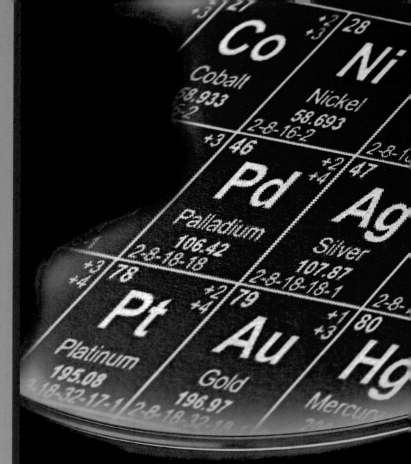

IB		IIB			
29	$+1$ $+2$	30	$+2$		
Cu		**Zn**			
pper		Zinc		Gallium	
46		65.39		69.723	
		2-8-18-2		2-8-18-3	
$+1$	48		$+2$	49	
Cd			**In**		
dmium			Indium		
2.41			114.82		
8-2			2-8-18-18-		
$+1$	81				
$+2$					

The atomic number of an element is equal to the number of protons in each atom, and defines the element. Each element has a unique atomic number.

An important way of learning and understanding elements for chemistry is the periodic table.

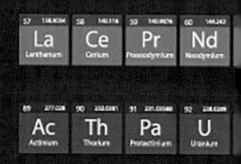

Periodic Table Of Elements

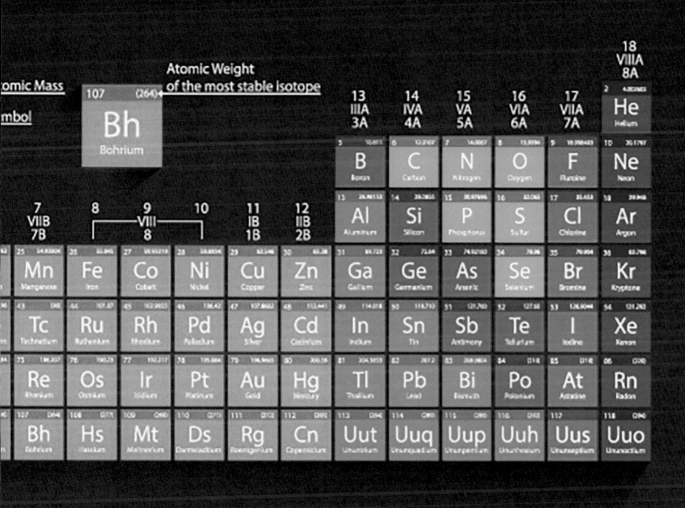

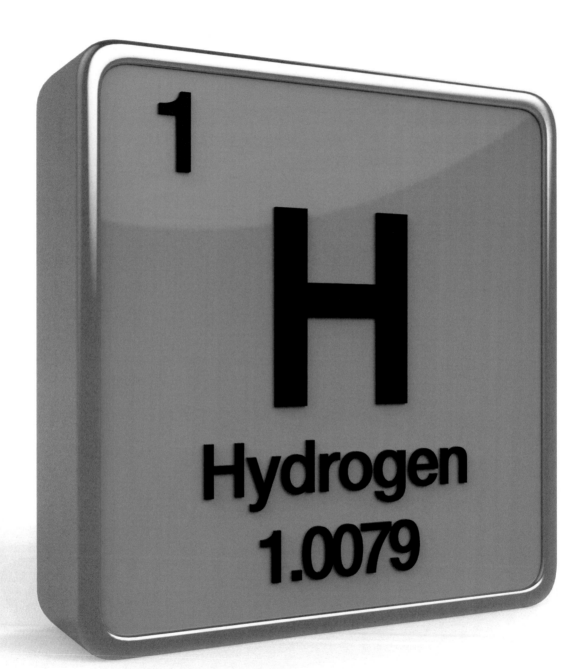

Hydrogen is the most common element found in the universe.

Elements found on Earth and Mars are exactly the same.

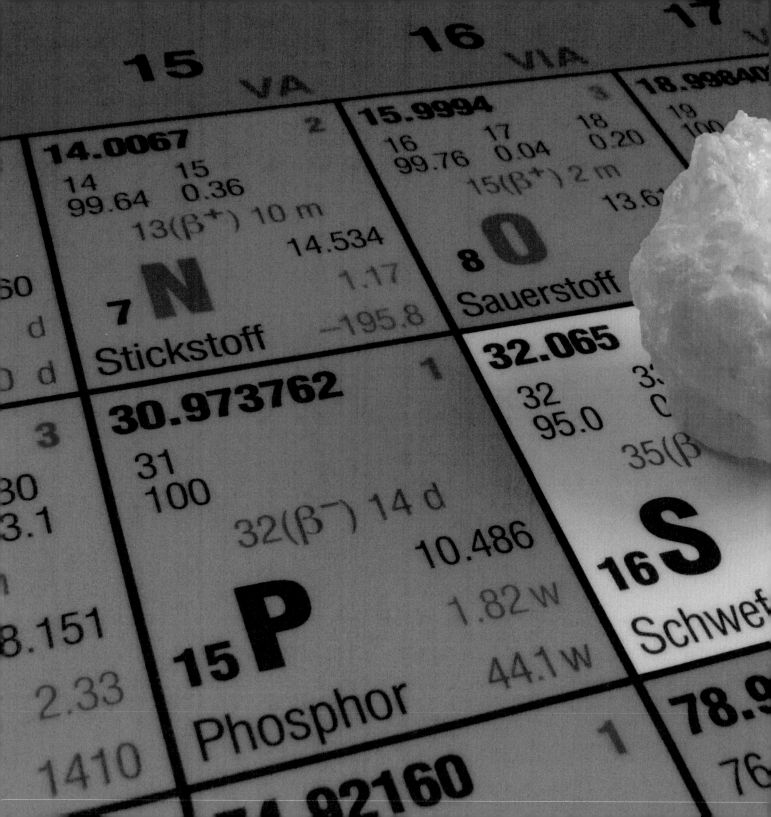

The known elements have atomic numbers from I through II8, conventionally presented as Arabic numerals.

Chemical elements are named after various things. Sometimes it is based on the person who discovered it, or the place it was discovered.

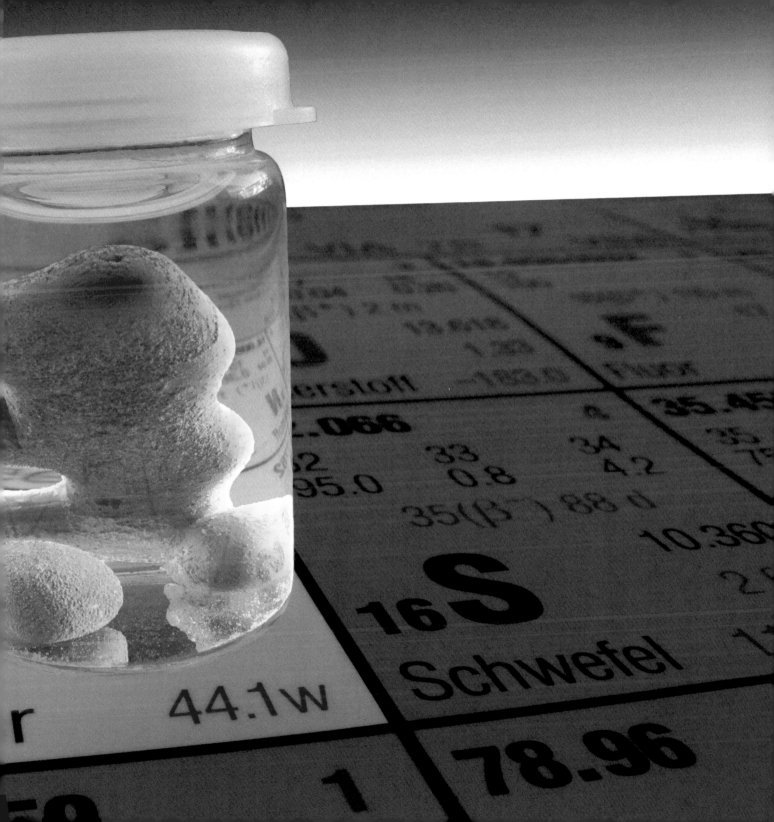